# English medieval cometary references over a thousand years

**By the same authors**

**By Austin Mardon** A Conspectus of the Contribution of Herodotus to the Development of Geographical Thought 1990; International Law and Space Rescue Systems 1991; Kensington Stone and Other Essays 1991; A Transient in Whirl 1991; Alone Against the Revolution 1996; Political Networks in Alberta 1905-1992 2002; 7 days in Moscow 2005; The Contribution of Geography to the Recovery of Antarctic Meteorites 2005;

**By Ernest Mardon** Narrative Unity of the Cursor Mundi 1967; The Founding Faculty of the University of Lethbridge 1968; Place Names of Southern Alberta 1970; The Conflict Between the Individual & Society in the Plays of James Bridie 1971; Who's Who in Federal Politics from Alberta Ridings 1972; English Studies at Canadian Universities 1972; Community Names of Alberta 1975

**By Austin Mardon & Ernest Mardon**

Alberta Judicial Biographical Dictionary 1990; Alberta Ethnic Mormon Politicians 1991; Alberta Ethnic German Politicians 1991; When Kitty Met the Ghost 1991; Down & Out & On the Run in Moscow 1991; The Girl Who Could Walk Through Walls 1991; Alberta Mormon Politicians 1992; Alberta General Election Returns & Subsequent Byelections 1882-1992 1993; Edmonton Political Biography 1994; Alberta Political Biographical Dictionary 1994; Alberta Executive Council 1905-1990 1994; Early Christian Saints 1997; Later Christian Saints for Children 1997; Many Christian Saints for Children 1997; Childhood Memories & Legends of Christmas Past 1998; Community Names of Alberta 1999; Men of Dawn 1999; United Farmers of Alberta 1999; The Genealogy of the Mardon Family 2000; Alberta Catholic Politicians 2000; Alberta Anglican Politicians 2001; Liberal Politicians in Alberta 1905-1992 2002; What's in a Name? 2002; Edmonton Members of the Legislature 2004; Senators and Members of the House of Commons from Edmonton 2004; Edmonton Municipal Politicians 2005; Alberta Francophone Politicians 2007

# English medieval cometary references over a thousand years

Compiled By

ERNEST GEORGE MARDON

AUSTIN ALBERT MARDON
&
CORA HERRICK

PUBLISHED BY GOLDEN METEORITE PRESS
Edmonton, Canada

A Golden Meteorite Press Book.


Printed in Canada.

Cover art by Andrew Phillpotts.
Published by Golden Meteorite Press.
Golden Meteorite Press.
Post Office Box 1223, Station Main,
Edmonton, Alberta, Canada. T5B 2W4

ISBN 978-1-897472-00-2

**Library and Archives Canada Cataloguing in Publication:**

Mardon, Austin A. (Austin Albert)
English medieval cometary references over a thousand years / Austin Albert Mardon, Ernest George Mardon, Cora Herrick.

ISBN 978-1-897472-00-2

1. Comets--Observations--History--To 1500.
2. Comets--
History--Chronology. 3. Astronomy, Medieval--England--History.
4. Cosmology, Medieval--England--History. I. Mardon, Ernest G.,
1928- II. Herrick, Cora, 1989- III. Title.

QB724.M37 2007 523.60942'0902
C2007-906966-5

**Dedicated to**

Catherine Mardon for entering the Mardon's life and enriching our collective lives.

**Acknowledgements:**

I want to thank our beloved wives for pushing me when I need it, and catching me when I fall. Thanks to the Lethbridge Public Library and the University of Alberta Library Rutherford branch. Thanks to Diane Herrick for her support. Thanks to Andy P. for the cover and to our many friends and supporters. And finally to generations of monks and learned men in England for the 1000 years that they recorded these comet events in their ancient skies.

**Preface:**

This book covers the citations of comets that were recorded in historical manuscripts during the Middle Ages by English chroniclers and other historic references. It is the authors understanding that as with any work of this type their might be the possibility that a historical source might have been missed. Since the time of Halley there has been much interest in historical comets. It was through historic records that the periodicity of comets was first proposed by Halley the great English astronomer. After viewing the large number of historic sources to cometary events in the English record it becomes obvious that comets as likely with other astronomical phenomena were commonly know of and held some significance. Historic records exist in India and China from this same period of time of comets and can when cross reference confirm the veracity of eachs observation. References to comets played important social importance as harbingers of doom or at least change.

Finally comets were perceived to be important enough to be recorded at the time during the Medieval period and give us a glimpse to those writers who looked up with awe at the sky.

## TABLE OF CONTENTS

1066 AD Anglo-Saxon Chronile: Winchester & Petersburgh, Matthew of Paris, Le Livere de Reis de Engletere, Geoffrey Gaimar Destorie iles Engles
1096 Anglo-Saxon Chronicle:Peterborough
1097 Henry of Huntington, Le Livere de Reis de Engletere, Cronica Rogeri de Hoveden, Anglo-Saxon Chronicle: Petersborough, MS
1104 AD Flores, Matthew of Paris History of England
1105 AD Anglo-Saxon Chronicle: Peterborough, MS
1106 AD Flores, Henry of Huntingdon, Anglo Saxon Chronicle: Peterborough MS, Cronica Rogeri de Hoveden
1109 AD Historia Novorum in Auglia, Cronica Rogeri de Hoveden
1110 AD Flores, Anglo-Saxon Chronicle: Peterborough MS, Cronica Angle Matthew of Paris, Henry of Huntington
1114 AD Anglo-Saxon Chronicle: Peterborough MS, Henry of Huntington, Matthew of Paris
1135 AD Le Livere de Reis de Engletere
1145 AD Sketch Eadwines Psalter
1146 AD Matthew of Paris, Flores Historarum
1165 AD Cronica Rogeri de Hoveden
1188 AD Annuals Combrae
1198 AD Yeoman's Only England Comet
1223 AD Flores, Matthew of Paris History of England
1240 AD Flores, Matthew of Paris History of England
1264 AD Willelmus Abbs Nonus
1274 AD Flores
1304 AD Painting Grotto

1315 AD  Flores Historiarum
1337 AD  Chronium a Monacho Sant, Albani, Thomas Walsingham Historia Anglicana, Cronicon Anglice
1368 AD  Thomas Walsingham Historiarum Anglicana, Cronicon Angliae, John Capgreve, Chronicon a Monarch Saints Albaril
1382 AD  Chronicles and Memorials of Great Britain (Volume 64) , Thomas Walsingham Historia Anglicana, Cronicon Angliae (Volume 64)
1401 AD Annuals Henri Quarti, John Capgrave
1402 AD Thomas Walsingham Historia Anglicana

This study of written confirmation of the appearance of comets extend from 497 AD to 1402 AD.  In this 900 year period a total of some fifty different comets were observed and that fact was recorded in the various chronicles of the time.

**497 AD**

Matthew of Paris Flores Historiarus
This is the first observation in Medieval England.

Flores

Historiarum

His ita gestis, apparuit stella mira magnitudinis et claritatis, uno tantum radio contenta, ad cujus radium erat globus igneus procedebant, quorum unus longitudinem suam ultra Gllicana climata extendere videbatur, alter vero versus mare Hybernicum vergens, in 1 vii radios minores terminabatur. Hoc sidere viso, Uther, nimio terrore percussus, sapientes suos sollicitabat, ut, quid stella prtenderet, indicarent; qui respondentes dixerunt, "Te quidem sidus et igneus draco "sub sidere significant; radius antem, qui versus Gallicanam"

**541 AD**

Yeomans states that this comet was observed in the Orient. It is the second appearance of a comet in Medieval England.

Flores

Historiarum

In Gallia cometa apparuit, ita quod caellum ardere videbatur. Eodem etiam anno verus sanguis ex nubibus efluxit;

**575 AD**

Chambers states that this possibly was a "daylight comet". The object seen in April may have been a NOVA.

**677 AD**

Ethelwerds Chronicle

Flores

Historiarum

Sub his quoque diebus cometa per tres menses apparuit, et unoquoque die mane velut sol resplenduit.

**678 AD**

Anglo-Saxon Chronicle: Parker Ms

Here the star comet appeared.

Ta. McJ Sevanton (London, Dent) p. 38

Anglo-Saxon Chronicle: Land Ms

Here the star comet appeared in August, and every morning for 3 months shown like a sunbeam.

Tr. M.J. Swanton, London (Rent) p. 38

**684 AD**

Cometa

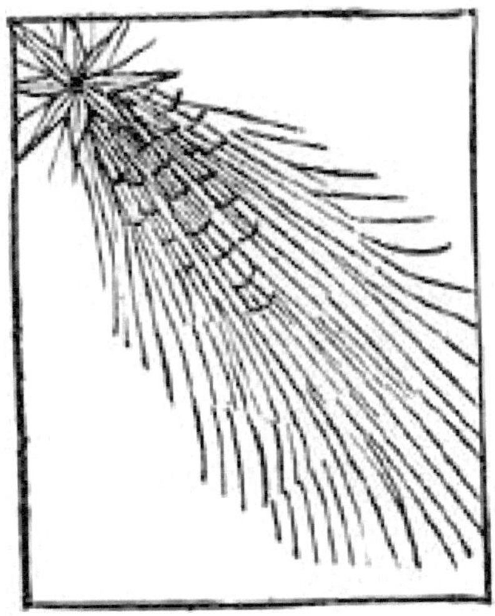

## 729 AD

Bede's Ecliastical History of the English People

Reteowdon twegen steorran ymb pa sunnan utan, pa syndon on bocum cometa nemde;

After this, in the 729th year from our Lord's incarnation, there appeared two stars about the sun, which in books are called comets; and they caused much alarm to the persons who saw and beheld them. The one of them preceded the sun in the morning, when rising, the other followed the sun in the evening, when setting, as if they were prophets of dire disaster at sunrise as well as sunset, betokening the coming of much ill to men at both times both by day and by night. A beam stood out from them, like a torch of fire, towards the north of the earth. These stars appeared in the month of January and remained two weeks. At that season, soon after, the most grievous pest, the Saracens, wasted and destroyed the realm of Gaul with grievous and miserable carnage; but they soon after received and suffered the due punishment for their perfidy in that same province.

Sr. Thomas Miller, Liber
Quintas, chapter 22, p. 477

Anglo-Saxon Chronicle: Parker, Ms

Here the star comet showed itself; and St. Egbert passed away.

Anglo-Saxon Chronicle Peterborough, Ms - In this year appeared two comets.

Flores-

Historarium

Apparuerunt circa solem duae stellae horribiles, quarum una solem praecedebet orientem, et altera sequebatur occidentem, quasi Orienti siul et Occidente dirae cladis praesagae.
Henry of Huntington

Edelhardi regis anno tertio apparuerunt dose cometae circa solem terribiles; una quipped solem praecedebat... orientem, alia sequebatur occidentem, quasi Orienti simul et Occidenti dirae cladis praesagae; vel certe una diei, altera noctis paecurrebat exortum, ut utroque temprore mala mortalibus imminere signarent. Portabant autem faciem ignis contra Aquilonem quasi ad accendendum acclinem; apparebatque mense Juanuario, et duabus ferme septimanis permanebant. Quo tempore gravissinia Sarracenorum lues Gallias et Hispanias late misera caede vastabat; et ipsi post non multo in eadem provincia dignas suae perfiiae poenas luebant.

Matthew of Paris

Chronica Majora

Apparuerunt circa solem duae stellae Hen. Hunt. Horribiles, quarum una solem praecedebat orientem, et altera sequebatur occidentem, quasi Orienti simul et Occidenti dirae cladis praesagae; vel certe una diei, altera noctis, praecurrebat exortum, ut utroque tempore mala mortalibus imminere signarent. Portabant autum fascem ignis contra aquilonem ad occidentem acclinem, apparebatque mense Januario, et fere per dies quindecim permanebant. Quo tempore gravissima paganorum lues Gallias et Hispaniam miserabili caede vastabant, sed et ispi non multo post in eadem provincia dignas suay perfieiae poenas mirabiliter luebant.

Bartholemew of Cotton

Apparuerunt dual Cometa

This comet, referred to as the Venerable Bede's Comet, is the most well documented of the early Medieval observations of a comet. A total of eight chronicles mention it. Bede himself took it as a sign of doom as the Islam host crossed from Spain to Southern France. The invading armies were defeated by Charles Martal, grandfather of Carla Magna at the Battle of Tours in 1732.

**793 AD**

Anglo Saxon Chronicle: Parker, Ms.

Here terrible portents came about in the land of Northumbria, and miserably afflicted the people: these were immense flashes of lightning and fiery dragons were seen flying in the air.

Here terrible portents came about over the land of Northumbria, and miserably frightened the people: these were immense flashes of lightning, and fiery dragons were seen flying in the air.

## 868 AD

Flores Historiarum

“Tunc Cometa Apparuit Manifeste”

**891 AD**

Flores Historiarum

Eodem anno circa tempus rogationum coneta apparruit, quae anglice verede sterpe nuncupatur.

**892 AD**

Anglo Saxon Chronicle- Parker, Ms

And the same year after Easter during Rogation-tide or earlier appeared the star which in Latin is called “comet”, likewise men say in English that a comet is a (flax) long-haired star, because long beams of light shine there forth, sometimes on one side, sometimes on every side.

Ethelwerd Chronicle

## 901 AD

Bartholomew of Cotton
Apparuit Cometa

**905 AD**

Anglo Saxon Chronicle

Here a comet appeared.

Anglo Saxon Chronicle : Worchester, Ms

In this year the comet appeared thirteen days before the Kalends of November 1.

Henry of Huntington

Secundo ab hoc anno luna valde obscurata est, et magno fuit horrori vidintibus: tertio anno apparuit cometa:

**906 AD**

Flores Historarium

Cometa apparuit fere per dimidium et occisionem, quae Paulo post in praelio inter Danos et Anglos dampnose commissa est, in quo utriusque gentis tot nobiles corruerunt

Chronicon Angliae Petriburgen

Comet Visus est

The author of this chronicle belongs to the group of Post Norman Conquest of England Chroniclers who wrote not in Old English but in Latin. The first entry in this work is dated 604 AD and the last is 1368 AD.

**915 AD**

Bartholomew of Cotton

"Comet Apparuit"

[In this year a comet appeared]

**971 AD**

"Annales de Waverlesa"

cujus regni principio apparuit cometa, signum maximae famis quae anno secundo secuta est. Rex Edgarus sepultus est Glastoniae

**975 AD**

Anglo Saxon Chronicle: Land, Ms.

And in that year Edward, Edgar's son, succeeded to the kingdom; and then immediately in harvest time in the same year, the star comet appeared, and then in the following year came a very great famine.

Anglo Saxon Chronicle: Worchester, Ms.

And in this year Edward, Edgar's son, succeeded to the kingdom, and soon at harvest time of the same year appeared that star known as Comet. And the next year came great hunger

Henry of Huntington

Edwardus filius Edgari Regis, qui dictur Sanctus, regnavit pro patre suo: cujos regni principio apparuit cometa, signum scilicet maximae famis quae in anno sequente secuta est.

Chronica Johannes Oxenedes

Tune visa est cometes, qui vel peste provincialum vel regni mutationem portendere pro vero asseveratur

Bartholomew of Cotton

Apparuit cometa

John of Oxford

Tune visa est cometes, qui vel pestem provincialium vel regui

Cronican Angliae Petriburgen

Comet Apparuit

And then up in the heavens appeared a star in the firmament which heroes, firm in spirit, prudent in mind, men learned of science, wise soothsayers, widely call by the name of comet. The Ruler's vengeance was widely known, famine over the earth throughout the nation of men.

Here Edward, Edgar's son, succeeded to the kingdom, and then immediately in harvest time in that same year, the star comet appeared. And then in the following year came a great famine and very manifold disturbances throughout the English race.

**976 AD**

Flores historiarum

Cometa Apparuit, et dira fames subsecuta est.

## 977 AD

Le Livere de Reis de Engletere

A cel hure vist hom une esteile, ki hom apele commete, ki signefie pestilence, hu change del reaume, e mort de rei. Kar tost a pres la tere devint baraigne, e ne porta gueres de fruit, don't grant famine out par tut en lepais, e les bestes morurent com a la gleive. Tut le reaume ausi de tote parz fust truble; kar par tute la Marche furunt moignes de lur eglises engetez,

At that time there was seen a star called a comet, which signifies pestilence or revolution of government or death of the king. For soon after the land became barren, and bore hardly any increase, whence there was great famine throughout all the country, and the cattle died as by the sword. All the kingdom also on all sides was troubled; for throughout Mercia the monks were cast out of their churches,

**995 AD**

Anglo Saxon Chronicle: Peterborough, Ms.

In that year the star comet appeared; and archbishop Sigeric passed away.

Anglo Saxon Chronicle: Canterbury, Ms.

**1005 AD**

Annales de Waverleia

Cometes horribilis in austro visus est

**1066 AD**

Anglo Saxon Chronicle: Winchester, Ms.

In this year King Harold came from York to Westminster at the Easter which was after the midwinter that the king passed away; and Easter was then on 16 April. Then throughout all England, a sign such as men never saw before was seen in the heavens. Some men declared that it was the star comet, which some men call the 'haired' star; and it appeared first on the eve of the Greater Litany, that is on 24 April, and shone thus all week.

Anglo Saxon Chronicle: Land, Ms.

Mathali Paris

Apparuit in Normannia cometa mense Augusto, maximum cernentibus incutiens timorem, radiosque suos versus Aangliam extendens; eodem tempore Willelmus Comes, congregato exercitu classeque parata, circa festivitatem Sancti Michaelis ventis vela committens, in Angliam transvehitur, nulloque resistente, inter duo castra, Hastinges atque Pevenesellum, terram Anglorum ingreditur, cui rei astipulantur versus illi quos dominum Lanfrancum eo tempore ferunt fecisse. Sunt autem hii:

"Anno milleno sexageno quoque seno"
"Anglorum mete flammas sensere cometae"

Le Livere de Reis de Engletere

Harald, le counte de Kent, le fiz le counte Godwin, tantost len demein, cest adire le jur del Epiphanie, prist la corune par sa auctorite demeine, e fust coroune de Aldred le erceveske de Euirwik a Loundres, e ne regna pas un an. Memes le an une esteile, ki hom apele comete, fust vewe non pas soulement en Engletere mes par tut le mund, les viii. Kal. De Mai; e resplendist de trop grant claret.

Harold, the earl of Kent, the son of earl Godwin, soon, on the morrow, that is to say, on the Epiphany, seized the crown on his own authority, and was crowned by Aldred, the Archbishop of York, at London, and reigned not so much as a year. The same year a star, which they call a comet, was seen, not only in England, but throughout the whole world, on April 24, and shown with very great splendor.

Henry of Huntingdon

Et facta cst mutatio dexterae Exeelsi; quam cometa ingens in exordio ejusdem anni dosignaverat. Undo dietum est:

"Anno milleno sexageno quoquue seno"

"Anglorum metae flammas sensere cometae'

Chronica Rogeri de Hoveden

Eodem anno, viii. Kalendas Maii, stella cometis, uon solum in Anglia, sed etiam, ut fertur, per totum mundum isa, per vii. Dies splendore nimio falgebat. Unde quidam sic ait,

Anno milleno, sexageno, quoque seno,
Anglorum metae crinem sensere cometae.

Flores Historiarum

Hane autem regni subversionem sanguinea atque crinita, in exordio illius anni apparens, minaci fulgore praesignavit, unde quidam,

Anno milleno sexageno quoque seno,
Anglorum metae flammas sensere cometae.

Apparuit in Normannia cometa mense Augusto, maximum cernentibus incutiens timorem, radiosque suos versus Angliam extendens; eodem tempore Willemus Comes, Comgregato exercitu classeque parata, circa festivitatem Sancti Michaelis ventis vela committens, in Angliam transvehitur, nulloque resistant,e, inter duo castra, Hastinges atque Pevenesellum, terram Anglorum ingreditur, cui rei astipulantur versus illi quos dominum Lanfrancum eo tempore ferunt fecisse. Sunt autem hii:

"Anno milleno sexageno quoque seno

"Anglorum mete flammas sensere cometae."
Geoffroy Gaimar  Lestorie des Engles

After their death a comet
(A star, that is of which soothsayers and good astronomers know that it portends either good or ill) showed itself in the firmament. Many people saw it well. On the night of "Litania Major" it made as much brightness as if it were day. Many men looked at it. In many places they foretold from it. Each man said his guess, but soon followed the great strife, and the great tribulation, which afterwards came to the country. Then came Tostig with much folk.

Liber de Hyda

A star which prophets, and clerks learned in astronomy, when it shows itself, know well that God makes this sign seen, that the people will know that there are some who do none of all his commandments; That his faithful people have not peace among themselves,

Annales de Morgan

Comet Apparuit

[In that year a comet appeared]

Annales de Waverleia

Tune gens hujus patriae pecuniam et obsides illi dederant, et terram suam erga cum

relevaverunt:
Anno milleno sexageno quoque sene,
Anglorum metae flammas sensere cometae

Quam cometa ingrens in exordio ejusdem anni designaverat: unde dietum est,
How anno seno milleno bisque triceno,
Anglorum gentes bello periere cadentes,
Haroldus cecidit, Willelmus dux superavit.

Draco Normaisus

Hujus temporibus resplenduit ipsa cometa,
Cujus ab igne novo Neustria clara nitet:

**1067 AD**

Geoffrey Gaimar

Lestorie des Pingles

Many folk saw a sign
In likeness of fire it was,
In the air it greatly flamed and burned:
Towards the earth it approached,
For a little it quite lighted it up.
Then it revolved up above,
Then fell into the deep sea.
In many places it burnt woods and plains.
There was no man who was certain
Nor who knew what this meant,
Nor what the sign portended.
In the country of Northumberland
This fire went about showing itself;
And in one year, in two seasons,
Were these displays.

Chronicles of the Reig of Stephen, Henry II and Richard I

Draco Normanus

Anno quo moritur tria signa stupenda coruscant, Portentum superum, sed latet istud adhue. Parisius juxta metuenda cometa refulsit, Martius hane fudit, belliger ipse deus. Circulus aethereus solem circa radiavit Maximus.

Quae tamen haee extent dicere nemo valet.

Jura novi regni forsan canit ipsa cometes,
Luna rubens belli tristia fata refert.

**1077 AD**

Old English records tell of a "blazing starre seen near unto the sonne" on Palm Sunday. Prigre suggested that this may have been Venus near inferior conjunction, but the term "blazing star" was often used as a phrase to describe a comet.

**1087 AD**

Eulogium

Patris etiam memoriam ingenti congerie auri et argenti cum gemmarum luce conspicue adornavit.

Sexagenus erat sextus millesimus annus, cum pereunt Angli stella minante cometa.

Anno millesimo sexageno quoque seno Anglorum metae crinen sensere cometae.

**1096 AD**

Anglo Saxon Chronicle: Peterborough, Ms.

**1097 AD**

Henrici Huntendunensis

Eodem anno cometa apparuit.

Le Livere De Reis De Engletere

En le tenz cesti rey muz des doleruses aventures vindrunt en Engletere, sicom terremote, grant noyse de toneire, vent a desmesure, cretine de eawe, famin e murine des genz. Une esteile ausi, kom apele commete, fust veue XV jusr. Hom poeit ausi sovent veer ie dyable apertement, ke il parla oue genz ki passserent en boys e en forez e la u nul commun passage fust.

In the time of this king many grievous disasters befell in England, such as an earthquake, much noise of thunder, excessive wind, floods of water, famine and murrain among the people; a star also, which they call a comet, was seen for ten days; moreover, people could see the devil openly, as he spoke with persons who were passing in woods and in forests and in places where there was no common way.

Cometa III Kalendas Octobris Per XV Dies Apparuit

Cometa iii kalendas Octobris per XV dies apparuit.

Anglo Saxon Chronicle: Parker, Ms.

Da upon sce Michaeles maessan III n Octobr aetywde an selcuth steorra on aefen scynende sona to setle ganngaende. He waes ge sewen suth weast se leoma de him ofstod waes stithe lang ge duht suth east scinende fro neah ealle da wucan on das wisan aetwde. Manige Ven leto da hit cometa weare.

Then after Michaelmas fourth day before the Nomen October there appeared a rare star shining in the evening, and soon sinking into its setting. It was seen in the southwest, and the beam of light which stood out from it seemed very lone, shining in the southeast, nearly all week it appeared in this way. Many men supposed it was a comet.

Annades Waverleia

Cannales Monastic

Postea vero, iv, non. Octobris, apparuit quaedam stella mirabilis, lucens in vespere et statim se abscondens, et visa est Sudest, et claritas longa ab ila exiens lucebat in Sudest, et sic tota hebdomada visa est, et dicebatur cometa a quibusdam.

Chronicon Thomae Wykes

Cometa apparuit

Annales de Bermardeseia

Hocanno Kalendis Octobris apparuit Cometes XV diebus

Eulogium

Deeino anno kalendas Octobris apparuit cometes XV, diebus majorem crinim emmittens and Orientem, minorem versus Austrum. Apparuerunt et aliae stellae quasi jacula inter se emittentes. Ille fuit annus quo Anselmus lux Angliae, ultro tenebras erroneorum effagiens, Romam ivit.

Annales de Waverleia
Annales Monastici

Postea ante fesum Sacti Michaelis, scilicet v. kal. Octobris, apparuit caelum tota nocte paene tanquam arderet.

Annales de Morgan

Sexto non Octobris stela cometes apparuit

**1104 AD**

Flores

Cometa apparuit mense Februarii

A comet appeared in mid February

Matthew of Paris:  History of England

Eodum anno apparuerunt circa solem iiii circuli mirabiliter picti in meridie.  Cometa quoque apparit, menst Februarii.

**1105 AD**

Anglo Saxon Chronicle: Peterborough, Ms

Comet

John Capgrave "Illustribus Henricis"

In hujus rei signum , in codem anno, cometa apparuit, anno seilcet Domini

This was the first of the 17 comets observed in the course of the 12th century.

Annales de Waverleia
Annales Monastici

Cometes horribilis in austro visus cst. Hilpericus librum de compoto hoe anno scripsit. Fames maxima in Anglia fuit, et tanta quantam nullus hominum me minisse potuit. Exercitus Damorum recessit ad patriam suam.

**1106 AD**

Anglo Saxon Chronicle: Land, Ms

In the first week of Lent, on the Friday, 16 February, an unusual star appeared in the evening, and for a long period after that was seen shining for a while each evening. The star appeared in the south-west; it seemed to be small and dark, but the light that stood from it was very bright and seemed like an enormous beam shining north-east; and one evening it was seen as if the beam were forking in the opposite direction, towards the star. Some said that they saw more unknown stars at the time, but we cannot write about it more clearly because we did not see it ourselves.

Flores Historiarum

Eodem anne, in hujus rei signum, viae sunt duae lunae in caelo, una in Orientee, alia in Occidente, et cometa uno quasi cubito distans a sole, circa horam tertiam. Hic quoque contitutus est ordo canonicorum in ecclesia sanctae Mariae de Suwerc Londoniis.

Matthew of Paris: History of England

Signum terribile. Signum terribilius.

In hujus rei signum eodem anno cometa

apparuerat, a sole distans quasi cubito uno, ab hora tertia usque ad horam nonam, radium ex se lonum emittens. Visae sunt etiam in Corna Domini lunae binae et plenae, una in oriente, et altera in occidente.

Chronica Rogeri de Hoveden.

I prima vero septimana Quadragesimae, feria vi. Xiiii. Kalendas Martii, in vespera ostensa est quaedam insolita stell, et per xxv. Dics, eodem modo, cadem hora visa est lucere inter austrum et Occidentum. Parva qukem visa est, et obscura, sed splendor qui de ea exivit valde erat clarus, et quasi ingens trabes de orientali et aquilonali parte claritas ingessit se eandem stellam. Quidam dixerunt se plures insolitas stellas vidisse illo tempore.

Annales de Morgan
Annales Monatics

Stella quae cometa vocatur apparuit xiv, kal. Martii, in diebus Regis Henrici primi Anglorum, in eadem nocte quae tune praecessit diem Veneris in parte destra Zephyri qui Affricus nominatur, id est inter meridiem et occidentem, quae nimis obscura et perva intuentubus esse videbatur; sed tamen coma ejus, id est radius, nimis longa et magna haabebatur; luna erat eodem die undecima; et xxiv. Diebus perduravit.

Annales de Waverleia
Annales Monastica

Tune in prima ebdomada Quadragesimae,

Sexta feria, in vespertino tempore apparuit quaedam stella insolita, et diupostea visa est lucere unoquoque die, vespere, apparebat autem stella in Suduest tanquam parva et obscura;

Annales de Wisitonia

Stella quae cometa apparuit per XXIV Noctes

This is among the comets in this Medieval period that Gruntin in 1870 listed as "remarkable" by virtue of their size, brilliance and generally spectacular features. The other comets on his list include the comets of 1066, 1264 and 1402.

**1108 AD**

Anglo Saxon Chronicle: Parker, Ms.

After that, in the month of June, a star appeared in the north-east, and its rays stood out in front of it to the south-west, and was seen thus for many nights; and later on in the night, after it climbed higher, it was seen going away to the north-west.

**1109 AD**

Historia Novorum In Anglia

Ipso anno apparuit stella quam cometam plurimi nominabant. Visa est autem in mense Decembri circa lacteum circulum, crinem in australem coeli dirigens plagam.

Chronica Rogeri de Hoveden

Stella Cometa vi idus jusii apparuit et per tres hebdomadas lucere est

**1110 AD**

Flores Historiarum

Hoc etiam anno cometa apparuit more insolito

The Peterborough Manuscript – Anglo Saxon Chronicle

Tre ow waestmas wurdon thaere nihte thurh forste swide for numene. Thaer aeter on junies monthe aetywde an steorra northan eastan his leoma stod to foran him on thet suthwest thus manega niht waes ge saewan furthor nihtes syththan he ufor astah he was ge sewan on baec on the north west ganende

"Tree-fruits were that night largely taken by frost. Thereafter in the month of June appeared a star in the northeast. Thus it was seen for many nights. Further on in the night, when it rose higher, it was seen going back in the northwest.

Matthaei Parisiensis Historia Anglorum

Hoc etiam anno cometa apparuit more insolito, nam cum ab oriente surgens in firmamentum ascendisset, non progredi sed regredi videbatur.

Henrici Huntendunensis

Hoc in anno apparuit quaedam cometa more

insolito.

Annales de Waverleia

Et in quinta nocte Maii mensis apparuit luna clara in vespere, et postea parum coepit obscurari, ita quod ex toto ex illa nihil apparuit usque parum ante diem, et tunc iterum clara et plena fuit. Fuitque illo die decima quarta, et aer tota nocte clarus erat, et stellae lucentes, et fructus arborum noete illa gelu periere. Postea in mense Junii apparuit quaedum stella in Nordest, et lumen ejus porrigebatur ante cam in Sudwest, et ita multis noctibus visa est, postquam in noctabus sursum ascendrat, visa est ire retro in Norduest.

Annales des Bernard esela

Et eodem anno, Inna extinata, cometa mense junio apparuit.

**1112 AD**

Annales de Waverleia

Visa est quaedam stella in fine Maii insolita, com longo lumine multis noctibus lucens.

**1113 AD**

Flores Historiarum

Eodem anno, mense maio, cometa ingens apparuit.

**1114 AD**

Anglo Saxon Chronicle: Peterborough, Ms.

"In this year records the end of May was seen a strange star with a long beam shining."

Henry of Hantingdon

Cometa ingens in fine. Maii apparuit.

Matthei Parisiensis Historia Anglorum.

Kalendas Maiii. Tempests eodem quoque anno diversae tempestates mortales exterruerunt, menseque Maii cometa apparuit.

Annales de Berrmundereia

Cometa etiane mense Maro apparuit.

**1115 AD**

Annales de Morgan

Cometa apparuit
[In this year a comet appeared}

**1135 AD**

Le Livere de Reis de Engletere

Ki fust le drein an de son regne, le steile ki lem apele commete aparust, e tant de vent fust par mi Engletere ke il abati les turs e les boys tut anent.

AD 1135, which was the last year of his reign, the star appeared which they call a comet, and there was such a storm of wind through England that it levelled the towers and forests quite to nothing.

Draco Normanus

Cernitur ex uno substituisse duos. Hic quantus fuerit vivens, in morte cometa. Praecinit innumeris irradiata comis.

## 1142 AD

John of Oxford

Cometa apparuit.

**1145 AD**

Eadwine Psalter, an illuminated-manuscript collection of the Psalms, was copied by the monk Eadwine from the earlier Ultrecht Psalter. Judging from Eadwine's dates, stylistic evidence and this representation of a comet at the bottom of a page, the copy was made in or soon after 1145, when Halley's Comet appeared. The legend refers to the radiance of the comet, or "hairy star", and remarks that comets appear rarely, 'and then as a portent'. Above the comet are three Latis versionsof Psalm 5: Romanum and Gallicanum.

## 1144 AD

Annales de Theokesberie

Cometa Apparuit

**1145 AD**

Annales de Bermundeseia

Halley's comet was observed by an English monk/scribe Edwin.

Annales de Wistonia

Hoc anno cometa apparuit.

**1146 AD**

Matthaei Parisiensis Historia Anglorum

Circa idem tempus cometa, nunquam boni signiticativa, diebus multis apparuit in occidente, circumquaque latissime coruscando.

Flores Historiarum

Eodem anno similiter cometa in occidente, circumquaque mirabiliter radians, apparuit

**1165 AD**

Chronica Rogeri de Hoveden

Eodem anno duae cometae apparuerunt ante solis ortum, mense Augusti; una ad Austrum, altera ad Aquilonem. Cometa est stella, quae non omni tempore, sed maxime in bitu Regis aut in excidio regionis apparet. Et quando com crinito diademate apparuerit fulgens, regale nunciat lethum: si autem ferens comas, rutilansque sparserit illas, patriae monstrat excidium.

**1188 AD**

Annales Combrae

Comete britanniae per totam insulam visus est

Annales Combrae

Cometes Britanniae per totam insulam visus est, nec sine re signavit enim Henrici Regis mortem, qui cito post obiit; cujus visio nunquam impune mortalibus habeteur.

**1197 AD**

Annales de Theokesberia

Cometa apparuit tota hyerae fere.

**1198 AD**

November: England

A comet appeared for 15 days.

Note: This is the first reference made by Donald K. Yeomans to England. At that time King Richard I was on the throne.

**1199 AD**

John of Oxford

Cometa per xv. Dies mense Novembri per diem apparuit, interitum Regis Ricardi prodigialiter, ut aiunt, praefigurans, sieut et hyemalia borrenda tonitrua, quae in crastino Sancti Thomae apostolic audita sunt.

**1222 AD**

Annales de Dunstaphia

... visa est quaedem stella tam in Francia quam in Comet. Anglia,... quae a plerisque cometa dicebatur; et horis ... es apparuit, et micantibus radiis in Aquilo... nensis notabiliter rutilabat.

Chronicles of the Reigns: Stephea, Henry I and Richard S.
Continuatio Chornici: William of New Umburgh

Hoc anno visa est cometa magnos habens radios. Item visi sunt dracones, eadem die eadem hora, per diversas provincias.

Annales de Waverleia

Stella magua et insolita apparuit in autumno.

**1223 AD**

Flores Historiarum

Cujus scilicet Phillippi mortem cometa ardens et crituita, Paulo ante apparens, evidenter indicavit. Ipsius igitur obitu divulgato, Henricus rex Anglorum misit archiepiscopum Cantuariensem cum tribus episcopis ad Lodowicum, jam in regem coronatum, constanter rogans, ut redderet sibi Nomanniam.

**1239 AD**

Matthaei Parisiensis Historia Anglorum

De quodam mirabili signo in aere.

Vigilia autem sancti Jacobi, in crepusculo, non adhue apparentibus stellis, com esset aer purissimus, apparuit, ad instar Luciferi, stella maxima ut facula, quae se jaculabatur per tractus aereos, quasi sagitta ignea, clarissima in capite, relinquens ignem et fumum post se cum moveretur. Quae etiam pluries sie apparuit; censebatur autem draco vel cometa.

**1240 AD**

Flores Historiarum

Et eodem tempore, per totum mensem Februarii apparuit cometa occidentali, emittens radium versus orientem.

Matthaei Parisiensis Historia Anglorum.

Apparuit Cometa

Sub ejusdem temporis processu, videlicet per totum mensem Februarii, tempore serotino, versus occidentem, quaedam fusca stella emittens radium versus orientem apparuit; quam, quia extra ordinem stellarum insolita emicuit, cometa fuisse dicebatur.

John of Oxford

Cometa apparuit.

Annales de Wigornia

Cometa visa est in Februario

**1260 AD**

Annales de kWintonia

Cometa apparuit.

**1264 AD**

Willelmus Abba Nonus

Cujus tempore, anno Domini 1264, cometes tam notabills apparuit qualem nullus tune vivens ante vidit. Ab oriente enim cum magno fulgore surgens usque ad medium hemspaerii versus occidentem comam perlucidam protrabebat.

This comet was observed and its actions were described by the Dominican Monk Aegidas who states it was the first seen in the evening and after sunset, after several a day it crossed the Sun and appeared in the morning.

Willelmus Abbas Nonus

Deined, anno Domini 1264, cometes tam notabilis apparuit qualem nullus tune vivens ante vidit. AB oriente enim cum magno fulgore surgens usque ad medium hemisphaerii versus occidentem comam perlucidam protrabebat.

William of Newburgh

Interea mense Julii stella, quae dicitur cometa, matutinis horis apparuit, emittens ex ne radios versus austrum, et circa festum sancti Michaelis disparuit.

Roger Bacon observed a comet in July, and

described it in detail.
Chinese records may be interpreted as implying visibility after sunrise for about a month.

**1265 AD**

William Rishanger

Hoc anno apparuit Cometa, tam notabilis, ut nullus tunc vivens viderit talem prius. AbOriente enim cum magno fulgore surgens, usque ad medium hemisphaerii, versus Occidentem, comam perlucide pertrahebat. Et licit in diversis mundi partibus multa forte significaverit, hoe tamen unum pro certo compertum est, utcum plusquam per tres menses duraverit, cadem nocte

**1273 AD**

Hoc anno, in Vigilia Sancti Nicholai, terrae-motus, fulgara, tonitrua, draco igneus, et cometa, Anglicos terruerunt.

**1274 AD**

Flores Historiarum

Et Cometa Anglicos Terruerunt

Annales de Wigornia

Draco agneuset cometa Anglicos terruerunt.

**1298 AD**

Annales de Wigornia
Annales Monastica

Nono kal. Octobris in Occidente cometa manifestissime videbatur, et rakdios ad Orientem clarissimos extendebat.

**1299 AD**

Peter of Limoges

He wrote a track on this comet that appeared in late January and lasted until March. It was of a moderate size, and had a long tail. It was dark blue.

Chronica et Annales

Hoc anno, in Vigilia Epiphaniae, factus est termmotus in Anglia, et in crepusculo diei illus, cometa quaedam apparuit in Septemtrionalem, mittens radios lateraliter versus Orientem, velut ignivomos, per tres dies post solis occasum; quod caedis magnae ibidem in proximo futurae fuit pronosticum.

**1301 AD**

Halley's Comet

It was observed in China and also in Iceland.

Close up of comet in the "Adoration of the Magi" shows how the artist applied tempera and gold pigments to the plastered wall in textured strokes approximating the luminiscent appearance of the coma and tail of the actual comet, which he must have observed carefully a few years earlier. He depicted the intensely bright centre of the coma, or head, with what appears to be an eight-pointed star, building up layers of pigment over the star to diffuse the image. Some pigment is lost, revealing the red adhesive by which it was applied to the plaster.
(Scientific American)

The Adoration of the Magi

In this remarkable masterpiece, the Star of Bethlehem is depicted as a comet, a symbol of hope. It is the first realistic picture of a comet, probably drawing it according to his memory of the recent appearances. This is the fourth and last record of Halley's Comet. The others were in 1066, 1145 and 1222.

## 1315 AD

Geoffrey of Meaux, French physician and astronomer, observed this comet from mid December through February 12 and because of its height above the horizon and brightness, it was visible day and night.

Flores Historiarum

Regnabat denique circa polum Articum ingens cometa, radios suos vicissim expandens per diersas noctis, horas, super quatuour plagas terrae apparens, terendo cursum suum a vigilia Natalis Domini usque ad tertium decumum diem sequentum, praesignans lustrando sequentia mala diversis mundi partibus alternatim profutura.

**1337 AD**

Thomae Wlasingham
Historia Anglicana

Circa Kalendas Augusti, id est, per quindenam praecedentem et quindenam sequentem, cometa apparuit, praesignans fortassis durum bellum futurum inter regna; quae partim radios emittebat circa Orientem, et partim versus Austrum.

Chronicon a Monacho Saint Albam

Hoc anno – circa Kalendas Augusti, id est per quindenam sequenten Cometa Apparuit.

**1338 AD**

Chronico Angica

Mense maya visa stella cometa inter part Settentrionales et occidentales, quae versus Francian radios suos parrexit.

Chronicon a Monacho Saint Marie

Hoc anno, circa kalendas Augusti, id est, per quindenam praecedentem et quindenam sequentum, cometa apparuit, praesignans fortassis dirum bellum futurum inter rena; quae partim radios emittebat versus orientum, et partim versis austrum.

## 1367 AD

Chronicon a Monacho Santi Albani

Mense Mayo, visa est stella cometa inter partes septentrionales, et occidentales, quae versus Franciam radios suos porrexit.

**1368 AD**

Thomae Walsingham
Historia Anglicana

Anno gratiae millesimo trecentesimo sexagesimo octavo, qui est annus regni Regis Edwardi, a Conquaestu Tertii, quadragesimus secundus, mense Martio visa est stella comata inter partes Septentrionales et Occidentales, quae versus Franciam radios suos porrexit.

John Capgrave
The Chronicle of England.

In the month of March, there appeared betwix the north and the west a sterre thei clepe comata, directing his bemes rite onto Frauns.

**1382 AD**

This is the last appearance of Halley's Comet in the Medieval Period. There is no written chronical mention of this comet's visitation.

Cometa

Eodem vero anno stella cometa appint in occidents anglius cuamper quindenam

Thomae Walsingham

Historia Angliana

Hoc anno, mense Decembri, stella cometa apparuit in Occidente, amplius quam per quindenam.

Chronicon a Monacho Santi Albani

Cometa
Eodem vero anno stella comata apparuit in occidente amplius quam per quindenam.

**1401 AD**

Annales Henrici Quarti

Cometa apparuit mense Martio, primo inter Corum et Septentrionem, videlicet, in Circio, flammas emittens terribiles, in altitudine magna porrecta, postremo comas in Boream transferens, in qua plaga ultimo videri deiit; designans, ut opinor, humanum sanguinem effundendum circa partes in quibus apparuit, Walliae, videlicet, et Northumbriae, ut dicemus.

John Capgrave
The Chronicle of England

In this same yere of this man appeared a sterre, whech thei clepe comata, betwix the west and the north, in the monthe of March, with a hie bem, whech bem bowed to the north.

With the commencement of the 15th century saw the waning of the Medieval period in England. Chronicles were not interested in this form of historical writing. The influence and scholarship connected with classical greek and Roman authors. It Italy the Renaissance of the Quatro Cento became all the rage.

**1402 AD**

Thomae Walsingham
Historia Anglicana

Anno gratiae millesimo quadringentesimo seeundo, cometa apparuit mense Martio, primo inter Corum et Septentrionem, videlicet in Circio, flammas emittens, postremo comas in Borcam transferens; praesignans fortassis effuionem humani sanguinis circa partes Walliae et Northumbriae expost futuram.

Fist Comet

Chambers states it may have been visible in daylight. This comet is regarded by many as one of the finest ever seen. Italian records speak of it as being visible in daylight for eight days, becoming visible well before sunset and remaining in the sky until after dark. Although there were no records to the effect, if ever there has been a comet capable of casting shadows, this would have been it.

## Chronicles

Anglo Saxon Chronicle Ed. with translation by B. Thorpe. 2 vols. London, 1861

Editions of the Anglo Saxon Chronicle:

The Parker Chronicle. Ed. by AH Smith. New York: Appleton-Century-Croft, 1966

The Parker Chronicles and Laws, Corpus Christi College, Cambridge, Ms. A facsimile. Ed. by R. Flower & H. Smith. Oxford Early English Text Study – Oxford Univ. Press, 1941

Peterborough Chronicle – Bodelian Manuscript Laud Misc. 636. Ed. by D. Whitelock Copenhagen: Rosen Kilde, 1954

Two of the 'Saxon Chronicles'. Ed. by John Earle. Oxford: Clarendon Press, 1865.

Two of the Saxon Chronicles Ed. by Charles Plummer. Oxford: Clarendon Press, 1889

The Parker Chronicle Ed. by AH Smith New York: Appleton-Century-Croft, 1966

Annales Cambriae (AD 444-1288) Ed. by J. Williams ab Ithel, M.A. London, 1860

Bacon, Roger Fr. Rogeri Bacon Opera quaedam hactenus inedita

1 Opus tertium

2 Opus minus
3 Compendium Pilosophiae
Ed. by JS Brewer London, 1859

Bartholomaeus de Cotton Historia Anglicana (AD 449-1298) Ed. by H. R. Luard London, 1859

Capgrave, John Chronicle of England Ed. by F.C. Hinseston London, 1858

Chronicon Angliae Ed. by EM Thompson London, 1874

Eadmer Eadmeri Historia Novorum in Anglia. Ed. from manuscripts in the Library of Corpus Christi College, Cambridge, by M. Rule. London, 1884

Gaimar, Geoffrey Lestorie des Engles soluma translacion Maistre Geffrei Gaimar. Ed. by TD Hardy& CT Martin. 2vols. London, 1888-1889

Henry of Huntingdon Henrici Huntendunensis Historia Anglorum. The History of the English, by Henry, Archdeacon of Huntingdon from BC 55 to Ad 1154. 8 vols. Ed. by T. Arnold London, 1879

Hisden, Ranulf Poly chronicon Ranulphi Higden, monach Cestrenis. Together with the English trans. Of John of Trevisa and of an unknown writer in 15th century. Ed. by C Babington (Vols. 1-2) and JR Lumby (Vols. 3-9) 9 vols. London, 1885-1886.

Hoveden, Roger of  Chronica Magirtri Rogeri de Hovedene Ed. by W. Stubbs 4 vols. London, 1868-1871

Johannes de Oyenedes Chronica  Johannes de Oyenedes  (to 1293)  Ed. by H. Ellis London, 1859

Knighton, Henry Chronicon Henrici Knighton Ed. by JR Lumby 2 vols.  London, 1889-1895

Mannyng, Robert  The Story of England by Robert Manning of Brunne.  Ed. from MSS. At Lambeth Palace and the Inner Temple by FJ Furnivall.  2 vols.  London, 1887

Matthew of Westminster Flores Historiarum per Matthaeum west monesteri ensem collecti, Ed. by HR Luard.  3 vols.  London, 1890

Paris, Matthew   Matthaei Parisienise monalhi Sancti Albani.  Historia Anglorum, sive, ut vulgo dicitur, Historia Minor Item ejusdem Abbreviato chronicorum Angliae.  Ed. by F. Madden.  3 vols.  London, 1866-1869

Peter of Ickham livere devers de Brittainee le livere de reis de Engleterre.  [From Brutus to 1274, with two continuations to 1326 and a trans.]  Ed. by John Glover  London, 1865

Trokelowe, John de Johannes de Troklowe, et Henrici de Blaneford Chronica et Annales 1259-1296, 1307-1324, 1392-1406  London, 1865

Walsingham, Thomas Ypodisma Neustriae London, 1876

Walter of Coventry Memoriale fratris Walteri de Coventriai: The Historical Colletions of Walter of Coventry. Ed. by W. Stubbs 2 vols. London, 1872-1873

Wauren, Jehan de A collection of the chronicles and ancient histories of Great Britain, now called England. Ed. & trans. By W. Hardy and ELCP Hardy. 3 vols. London, 1864-1891.

William of Malmesbury Willelmi malmesbiriensis monachi de gestis pontificum anglorum libri quinque. Ed. by NESA. Hamilton, London, 1870

Bibliography

Jervis, Jane L. 1985. Cometary theory in fifteenth-century Europe. Dordrecht, Netherlands: D. Reidel Publishing Co. P.209.

Kronk, Gary W. 1984. Comets, A Descriptive Catalog. Hillside, New Jersey: Enslow Publishers, p.331

Maclagan, E. The Bayeux Tapestry London: Penguin Books, 1953

Marsden, Brian G. 1989. A Catalog of Cometary Orbits. Cambridge, Massachusetts: Smithsonian Astrophysical Observatory.

Maunder, ASD 1934. Old cometary records. Observatory 57:278-281

Newton, Robert R. 1972 Medieval Chronicles and the Rotation of the Earth. Baltimore, Maryland: Johns Hopkins University Press. Appendix II

Pedersen, O. and Pihl, M. 1974. Early Physics and Astronomy. New York: American Elsevier.

Pingre, Alexandre Guy. 1783-1784. Cometographie ou traite historique et theorique des cometes. Paris. 2 volumes.

Stein, J. 1910. A newly discovered document on Halley's comet in 1066. Observatory

33:234-238.

Thorndike, Lynn. 1945. Peter of Limoges on the Comet of 1299, Isis 36:3-6

Thorndike, Lynn. 1950. Latin Treatises on Comets between 1238 and 1368 A.D. Chicago: University of Chicago Press. P. 274

www.ingramcontent.com/pod-product-compliance
Ingram Content Group UK Ltd.
Pitfield, Milton Keynes, MK11 3LW, UK
UKHW041141300926
15471UKWH00011B/88

9 781897 472002